T0338769

ENGINEERING SCIENCE ...
YOU KNOW MORE THAN YOU THINK

TALES FROM PLAYGROUNDS AND YOUR OWN BACKYARD

TOM CLIFFORD

Kids .... In playgrounds, kindergarten, romps in the woods, kitchens, etc...... experience and understand forces, and materials, properties, and interactions, and more..... all central to engineering science. These understandings are crucial to contributing and enjoying grown-up life. This little book identifies many of these learning experiences and describes how each experiences teaches the kids and later the grown-up about engineering.

Order this book online at www.trafford.com
or email orders@trafford.com

Most Trafford titles are also available at major online book retailers.

Print information available on the last page.

ISBN: 978-1-6987-0967-3 (sc)
ISBN: 978-1-6987-0969-7 (hc)
ISBN: 978-1-6987-0968-0 (e)

Library of Congress Control Number: 2021920580

Trafford rev. 10/13/2021

www.trafford.com
North America & international
toll-free: 844-688-6899 (USA & Canada)
fax: 812 355 4082

Contents

Introduction ix
Preface xi

Some Examples

Speed of Sound in Different Materials 1
Angular Momentum... 2
Doppler Effect 3
Sound Transmission 3
Acceleration... 4
Echos... 4
Sticky vs Slick Surfaces 5
Lift 6
Wind Chill... 7
Insulation 8
Peripheral Vision 8
Heat Transfer.. 9
Sonic Booms 9
Static Electricity 10
Tornados 10
Weather Front 11
Aircraft Aerodynamic Control......................................11
Cohesion, Adhesion, Soil Mechanics 12
Leadership 13
Breakage, Fatigue Cracking Strain Rate 13
Fishing Psychology 14
Plants Need Water 14
Gravity 15
Wind Resistance..15
Materials' Thermal Response 16
Microwave Oven Response 16

Water Draining Down..17
Angle of Repose...17
Deceleration 18
Targeting The Intersection Point...18
Flammability, and Thermal Mass...19
Fire-Redardant Materials 19
Organoleptics 20
Balance – Moment Arm 21
Triangular Brace...22
Rocket Thrust, psi..22
Spectrum..23
Mixed Colors 23
Stiction...24
"Kinesthetic Feedback" 25
Astronomy..26
Pendulums..26
Thrust Vector 27
Heat vs Temperature..28
Turbines 28
High Velocity Water...29
River Meanders..29
Music. Resonance..30
Balance. Sensors 31
Surfaces 32
Behavior, "pecking order" 32
Weather Effects..33
Phase Changes...34
Wettability...34
Optics, Clarity, and Magnification 35
Predator, Prey..36
Failure Analysis. Chain of Causes 37
Suspensions, Solutions 38
Expertise, Critical Thinking...39
Filtration..39
Nesting...40

Materials' Elasticity..40
Heating. Radiant, Convection, Conduction 41
Hydroplaning 42
Materials Strength. Strain Rate 42
Materials. Thermal Expansion 43
Use of Tools 44
Soap Bubbles 44
Balancing...45
Density..45
Exercise Activity..46
Planning..46
Materials "going bad"...47
Camouflage..48
Ballistics 49
Rust 50
Strength of Materials 50
Magnets...51
Territoriality..52
Experimentation..52
Gag Reflex...53
Tides and Waves 53
Taking Things Apart 54
Cuts and Bruises 54
Pets...54
Math..55
Later..55

Final Comments...57
Afterword 59

Introduction

This little essay is my attempt to convince non-technical folks that:

* they already understand basic principles of physics, science, and engineering,
* they learned these rudiments of engineering science early on,
* they use and encounter many of these engineering principles daily,
* they do not need formal training to understand and appreciate real-world applications and experiences,
* engineers and scientists just apply these principles ..... nothing mysterious, mostly only clerical..... you can understand and evaluate their comments and creations,
* proper academic engineering/science training is, however, often necessary to create new products and processes based on these principles

I'll discuss personal anecdotes and observations, most of which are familiar to many of us. These events arose in playgrounds, kindergarten, in games, in vacant lots, kitchens, and romps in the woods. Kids discover, understand, and use these concepts long before they encounter the corresponding arcane and off-putting scientific terminology. The structure of the book will be: Observation > engineering science concept > applications. I'll try to show how each little observation or activity demonstrates a principle or two. I'll try to explain the scientific or engineering principle behind each one, that can be described mathematically. Those concepts and understandings enable folks to appreciate, and even the efforts by engineers and scientists to create products like high-speed rail-road systems, interplanetary spacecraft, temperature-controlled residences, and more.

A possible market for this essay might be for K–12 science and STEM teachers, as well as parents, to encourage the students to more readily accept and apply the concepts. A good way to put a kid to sleep is to say "today we'll learn kinematics: "acceleration = force divided by mass $(A = F/M)$, and velocity = acceleration times time $(V=AT)$". More better to say "remember how hard it was to get that big wagon moving, but you kept pushing and it went faster and faster and pretty soon it was booking down the block?" Well that's the engineering principle that launches interplanetary space ships beyond orbit or gets a freight-train up to speed. Connecting personal experience to engineering applications fosters understanding.

Preface

I've spent a pleasant and very satisfying lifetime doing industrial engineering and science, nothing spectacular, embedded in a society of very nice outside folks, many of whom recoil from and dismiss "engineering" and "technical stuff" as somehow sinister or threatening. They felt they couldn't understand it, and it had no part of their life. Wrong! They understand and use it, and enjoy it more than they think. They just don't know, and probably feel they don't need to know, the names of the technical concepts. This essay is a series of anecdotes, hopefully to shed a bit of light, each one featuring some specific technical principle. but those are not exact and there is plenty overlap, and that's OK.... the world is a complex place..... observations and principles overlap.

Speed of Sound in Different Materials

We (the author and his siblings) were fortunate to grow up near railroad tracks. Leaving aside, for the moment, the worrisome dilemma of being on bikes half-way across the long bridge over the Arroyo Colorado, and hearing and worrying about the approaching train, we played with something else: When your buddy, way down the rails, taps the rails with a hammer or a rock, you hear and feel the sound thru the rail long before you hear it thru the air. Sound travels faster thru hi-density medium. The speed of sound in low-altitude air is ~760 mph... in water it's ~ 4X that in air; in steel it's ~13X that of air. More subtly: the speed of sound also varies with temperature and density Engineers use these relationships for studies of water temperature/density and depth, in aircraft and weapons guidance systems, and much more. You could guess correctly that there is zero sound in deep space

Angular Momentum

Playing in an office chair, the kind that can spin around, is lots of fun. You can spin, slowly, with your feet and hands outstretched; but when you retract your arms and legs, you spin faster. You stretch your legs out and you spin slower! One time long ago, several of my buddies were on a spinning merry-go-round with all our feet outstretched spun up to a respectable speed by a beefy local kid, then on a go-signal we all retracted our feet and scurried to the center of the merry-go-round. It spun up lots faster!

We had to really hang on (more about centripetal force later). What gives? Magic? No. It's conservation of angular momentum. A spinning disc with a large diameter has more angular momentum than a smaller disc of the same weight and speed. Shrink the big disc and it will spin faster. It is the principle that figure skaters use to do their spinning routines. It also keeps your toy gyroscopes functioning properly. Beyond all that, and of more fundamental significance, and of very little daily relevance, the astrophysicists and quantum mechanics folks talk about spinning neutron stars and Heisenberg uncertainty and quantized eigenvalues in elementary particles. But all that's not important to most of us. We can experience and do experiments in a very important engineering principle in local playgrounds, and try to cope with board-walk rides, by ourselves ... nothing mysterious.

Doppler Effect

We all hear "... the train a-coming, a-coming round the bend ...". As the train approaches, the whistle gets higher, as the train passes, pitch of the whistle drops. That's the Doppler effect.

Think of the train as issuing a string of waves ... that's a certain pitch. If that string of waves is being pushed toward you, that compresses the wave and the pitch is higher. As the train passes and the string of waves is stretched out, the perceived pitch is lower. That figures.

Sound Transmission

We grab one end of a garden hose and tell a little brother to listen carefully into the other end. But first we blow into the hose to drench him with the water that's still in the hose. Little brothers are fun. But when sound is constrained and manipulated (grand piano, certain music speaker systems, the Australian didgeridoo, louder foghorn under a dense fog bank), you know what's happening. Concert halls are designed to focus and transmits sounds properly. Engineers build "anechoic": chambers (walls covered by foam shapes, to eliminate echoes) within which they study sounds carefully. Kids puzzle why some spoken words sometimes become indistinct, until they learn about "white noise" .... overlapping random sounds at the same frequency and volume. But that suddenly makes sense, based on their experience.

Acceleration

We feel something when we get started pushing by a buddy in a wagon, or in a race car; we feel pushed back in our seats. We feel a "rush". That's acceleration. When something is being pushed, and nothing is stopping it, it goes faster and faster .... it accelerates. The harder the push and/or the lighter the object, the more the acceleration. Note that the longer this acceleration is applied the faster you go. The mathematical relationship is acceleration = force/mass $A = F/M$. Make the mass smaller (lighter) you get more acceleration. Make the force bigger (6 liter 400Hp turbocharged Chrysler hemi engine, for example) and you'll get more acceleration. Same with rockets and interstellar solar-wind-powered spacecraft. You felt it, you understand it, nothing mysterious.

Echos

We shout across the canyon or at a big building several blocks away, and we hear our voice come back. That's an echo. The sound bounces off the distant object and comes back to us. A good way to estimate the distance of that large flat surface is to time the echo's return. Sound travels 760 miles per hour in air. That's 760 miles in 3600 seconds...... or 1 mile in 5 seconds..... or 1/5 of a mile in 1 second. Let's say the echo returns in one second. Your shout had to travel to the wall and bounce back to you. That's ½ second going and ½ second coming back which means the flat surface is one half of 1/5 of a mile... or 1/10 of a mile away. Engineers and scientists and whales and bats and other creatures use this principle (echo-location) to judge distances. Incidentally, if you were really curious about the depth of the well and are good at math, and had a good ear and a real good timer, you could estimate the depth of a well.

Sticky vs Slick Surfaces

Very early on you feel different surfaces: sandpaper is rough; soapy glass is slick; the sticky side of adhesive tape is sticky. You've learned to adjust your handling of objects accordingly. This characteristic is very important in many applications. ASTM (American Society of Testing Materials) has developed standards for the slickness or roughness of surfaces like roadways, flooring, etc. Highway engineers try to manage traffic safety thru efforts to mitigate "aquaplaning", which describes the sudden transition of a nice dry grippy surface to a slick water-film surface in the rain. The math gets tricky, but you can follow all their discussions.

Also, kids see clearly that surfaces act differently when wet: newspapers seems to like it; they get grippy. Most plastic surfaces (Teflon, silicones, polyethylene, etc. they learn later) do not .... water runs right off. Same with paints, as kids learn later. The selection of paint and surface prep became important. Personal anecdote: long ago: restoring/painting a vintage sports-car. Final top coat paint kept bubbling off in ragged blisters, until I realized I had earlier de-dusted the primer coat with a "dry" rag that had previously been used to clean up after a silicone calk window job. Silicone hates wet. The silicone residue kicked off the paint! I had to sand everything down to bare shiny metal and start over! Lessons learned.

Lift

Blow up a balloon; feel it getting bigger and harder. Engineers lift huge weights up short distances by using a big flat bladder and inflating it. Think <u>pounds per square inch.</u> Suppose you inflated one of those 6ft X 6ft bladders to 10 psi. That lift bladder is 72 inches by 72 inches or 5184 sq. inches. Inflated to 10 psi (pounds per square inch), that means it will lift 51,840 pounds (more than 2 1/2 tons), Those bladders are not very thick, so the lift will not be very high, but high enough to get forks/dollies under the big load, lifted high enough to complete the lifting process. Any kid would understand that whole process, especially since kids have fielded helium-filled balloons, feeling the lift, and feeling how the balloon expands!

Wind Chill

Kids feel colder, and in fact are colder, in a stiff breeze. The wind sweeps away the boundary layer of air that's close to your body temperature. Here's some observations on specific sets of conditions:

at 40F: 30 m/hr wind feels like 28F

at 20F: 30 m/hr wind feels like 1F

at 0F: 30 m/hr wind feels like –28F and you really feel that!

Certainly other situations are different, but the same principles apply, and are understood by grown-ups. For instance, petrochemical refineries depend on maintaining those big columns and reactors and process units at a certain elevated temperature. A stiff cold wind causes very, very expensive havoc unless mitigating measures are applied. Trans-Arctic pipelines are protected by sophisticated heater systems to cope with the chilling effect of local storms. Clothes are selected and worn to deal with high cold winds.

Insulation

When it's cold, kids quickly learn to select and wear "warm clothes". The clothes do not provide warmth. They just contain <u>your</u> warmth. Same with insulation on houses, buildings, reactor columns in petro-chem refineries, and the whole length of the Alaskan pipeline transporting warm oil thru Alaskan winters. The insulation selected should match the application. T-shirts should be OK for most South Texas summers, many layers of wool are OK for most Alaska winters. <u>Insulation</u> is characterized by its "R-value". The higher the better. Examples include rock wool at 3.14, and foam-in-place-polyurethane at 6.25. It's not that simple but kids and we understand the concept. Insulation (preventing the transfer of heat or cold) is fundamental to life on earth, as well as for countless operations (cars, houses, industrial processing, cooking, clothing, etc.). You learn the concept very quickly.

Peripheral Vision

Kids see things "out-of-the-corner-of-your-eye": images and motions outside, to the right or left of our normal field of view. Often these things are fast-moving, in and out of our vision. The capability varies: some individuals' conditions, as they age, restrict it, leading to "tunnel vision". One important report concerns the capability of airplane dog-fighting fighters, in WW1 and WW2 conflicts; aviators required quick perception of enemy fighters approaching quickly from all angles and with lethal intent. Spot em' fast and take appropriate action fast. The bad guys rarely selected a straight line directly toward you; they came from anywhere, intent on putting you out of the picture. Successful pilots needed good and quick peripheral vision.

Heat Transfer

Bare-foot: step in a spot of cold water on the floor. Feels cold. Step on a not-cold penny on the floor. It feels cold but It's not cold. What gives? The heat transfer coefficient. Copper transfers heat like water does. Stepping on a cold wooden poker-chip feels like nothing. Stepping on a copper penny might feel cold. Wood has a much lower heat transfer coefficient. Stick a metal pipe or spoon into your bonfire and you feel the heat immediately.... do the same thing with a stick of wood and your end of the stick stays cool. Wood has a much lower heat transfer coefficient. You learn early on that most metals conduct heat readily, whereas glass, wood, and other organics mostly don't. They make good insulation.

Sonic Booms

If you were lucky you have heard "sonic booms", when a fast jet blazes overhead. What gives? As the plane reaches/exceeds the speed of sound in air (above Mach one (760mph, but depends on temperature and altitude), its shock wave sweeps along the ground when the jet overhead is traveling. You hear it as a local boom, but the shock-wave sound is actually moving across the land under the jet. Sometimes a double-boom emanating from nose and tail of the aircraft is heard. You are able to follow discussions on this topic, because you have heard sonic booms.

Static Electricity

You've seen socks stuck to items of clothing right out of the dryer. Shuffle across the rug ... and zap your buddy. Static electricity is different from current electricity. It happens and satisfies the difference between the charges on two surfaces. It is satisfied when you separate the surfaces, or when there is that brief zap spark. You understand why the gas stations encourage/demand that you ground your plastic gas can (on the ground!) before pumping in your gasoline: sparks have caused serious consequences! You understand meteorologists explaining lightning: motion within the storm builds up a substantial charge. The only way this charge can be eliminated is by grounding (lightning). Rug-shuffle is the storm cloud, your buddy is the ground, that zap is the lightning. Same, same.

Tornados

Air spinning ... you've seen those little "dust-devils"..... you've even been in some! Usually in the desert, it's seen as a tiny tornado. Something is pulling the air up and it spins to get there faster. Explanations of that complex phenomena make sense, because you have seen the air doing exactly that up close. Again, the math is complex. Storms generate big swirling cloud masses, wind shear, differences in humidity/density; all combine in a big swirling situation, which sometimes results in a tornado. You have seen them, but hopefully not experienced them close up, and when the meteorologist experts discuss tornados, you understand.

Weather Front

Kids have seen and were spooked about that line of dark clouds coming in from the North-West; and watched as the skies darken and the winds pick up, and usually there is rain and maybe hail in your area, and they must find shelter! When you see weather reports on TV, you know what they are talking about. Satellite sensors and imagery, and the massive computations required to model and predict global weather are improving, so long range weather predictions are becoming more accurate. You know what they're talking about. You've been there.

Aircraft Aerodynamic Control

Kids have tried shooting an arrow without feathers ... it just tumbles thru the air. Imagine a missile without tail-fins fired in air-to-air combat .... useless! All airplanes have tail structures (horizontal and vertical fins) called empennage. These serve to keep the plane stable and to provide directional control. Rockets without tail fins? an albatross without a tail? Nope! Simple concept, countless examples.

Cohesion, Adhesion, Soil Mechanics

Mud clods. Hands-full of sand or dust or dirt? Nope! You cannot throw that. That fist-full has limited <u>cohesion</u> ... it doesn't stick to itself very well. But wet dirt or wet clay? that sticks to itself. What's different?_That has good <u>cohesion</u>. Squeeze it into a ball, and chunk it at your buddy. That works. Create a highway road-bed using the available local wet dirt? You must pack it down first, using something like a "sheep's-foot-tamper/compactor". Earthquakes are very good at showing which hillside has good cohesion, or which undersea shelf slips out and triggers a tsunami. Soils engineers are very mindful of the cohesion of their soils. Tall buildings built on unconsolidated land-fill have collapsed when an earthquake, even a small one, jiggles the soil a bit to cause it to "liquify" and lose necessary cohesion. Big buildings need good solid foundations: rock or hard soil or deep pilings. Water into dirt makes mud, same sort of thing happens when you make cookies or concrete cement. On a more personal level; try to ease a label off of a container.... if the label comes off intact, it had good cohesion, and was held on by <u>adhesion</u>, which you overcame. If it shreds and some remains on the surface and some of it crumbles in your fingertips, the label had good <u>adhesion</u>, but bad <u>cohesion.</u> Hundreds of other examples and situations come to mind.

Leadership

You have watched flocks of small birds in flight. Suddenly, they all turn and swoop the same way simultaneously. They seem to have no leader; how does this spontaneity happen? You might have guessed, I think correctly, that all of the birds are identical, in sensors and behavior patterns; and when the situation changes, they all sense it instantly and they all instantly react the same way. You've seen minnows in a pond all split simultaneously when a threat appears. When snorkeling or Scuba diving you look up surrounded by a ball of hundreds of little fish, and they all suddenly disappear. They all got the message, simultaneously. Grown-ups evaluating politics and government actions behave the same way, for better or worse.

Breakage, Fatigue Cracking Strain Rate

You have tried to break something, and to get it to break you crank it back and forth ... back and forth.... Tiny cracks get bigger and then the item fails. This is <u>fatigue cracking,</u> and it's a bIg concern of engineers: from turbine blades in jet engines, to automobile suspensions, to bridge structure swaying gently in the wind for decades. Many materials strength specifications in industry include cyclic stress testing. Also strain rate: You learned about "strain rate" as a breakage factor: Remember "Silly Putty"? pull it apart slowly ... it stretches out. Pull it apart quickly (high strain rate) it snaps apart! Strain rate is such an important element in the strength of industrial materials, that ASTM specifies precise tensile test strain-rate parameters, in their certs and classifications of strength of materials.

Fishing Psychology

Bait, lure, location, presentation, rig, process. In your pond and river fishing experiences you learn exactly what the Tournament pros are talking about, and then later you apply these same skills and reasonings when you go deep-sea fishing for marlin or sails or tuna or dorado ... figure out what the fish want, then you do what you think will get 'em. Of course you release them, unless you want to take a few tuna to the local smoke-house, and bring home some smoked tuna to your wife. Just sayin'!

Plants Need Water

You have pulled weeds ... some weeds have deep roots; some have roots that spread out horizontally. Great big trees are not really different. Oak trees, in the plains and grasslands, develop deep tap roots growing down to try to reach a necessary water table. In contrast, redwoods in the foggy coast-side develop shallow horizontal root systems, because the available water is from fog and rare rain on the local forest surface; and there is nothing down below except rock. Plants of all sizes evolve to exhibit understandable characteristics, in order to get at the necessary available water. On a global scale, you understand that countries need water to grow their crops to feed their people, and you will understand that water will become more valuable than oil in many global situations, as global warming changes weather patterns.

Gravity

Of course, one of the first concepts we kids all learn is that things fall. We experienced this (a lot!) on purpose or accidentally. The examples when we were young, and applications and examples when we are grown-ups are way too numerous and obvious to describe here, but lessons learned all apply.

Wind Resistance

Gliders, feathers, dandelions in the wind ..... You know very early on about wind resistance. Drop a lead weight: it falls: thump; drop a feather and it takes forever to hit the ground. Racers and designers create streamlined cars and airplanes. Those fancy slick bicycle racing costumes and helmets that they wear are intended to minimize wind resistance. That's all hype and fashion, unless you really are clocking 50 mph+.... then wind resistance becomes a factor. Folks understand wind resistance and streamlining, based on their early experience.

Materials' Thermal Response

Kids learn early about plastics (more accurately called polymers): some polymers melt (thermoplastics), others just get hard and brittle (thermosets). This is basic to polymer engineering science. You notice that some polymers like wax (yes wax is a very low molecular weight polymer), will readily melt when warmed a bit. The "heat distortion temperature" is an important property of structural and packaging and household plastics. Also, for thermosets, the embrittlement and loss of strength over time at temperature is a critical attribute. Automobile radiator hoses can embrittle and fail over time. Most materials in industry and commerce are evaluated and selected based on their ability to retain their function over time and temperature. Kids understand this.

Microwave Oven Response

Some plastic containers melt in a microwave oven and make a mess; other containers are OK. That depends of their chemical makeup. If metal objects are accidently microwaved they make a big sparkly noisy dangerous event. Similarly: certain decorated glass or ceramic bowls/dishes are NOT appropriate for microwave use: their decorations are metallic ink. This microwave effect is an engineering science subject, and is the reason for those "Microwave Safe" signs you see on kitchenware.

Water Draining Down

You experienced this early on and frequently, in spilled drinks, in sprinkler play, in playing in creeks with little rapids. That fact was fundamental to Medieval engineers' water-wheel power, and is the basis of modern hydroelectric power systems, as well as to understanding flash-flooding from rainfall up in the water-shed, (typically a big area up in the mountains). Water seeping down from roof leaks causes damage in houses. Understandable.

Angle of Repose

The shape of the objects makes a difference: try to make a stack of marbles (no) and make a stack of wood chips. (yes) Big difference!... and you can see why. Also: <u>Wet vs dry</u> makes a difference: try to make a sand castle on the beach with that totally dry sand far away from the water. No go. Try the same with the wet sand at the water's edge. It works. Back to <u>angle-of-repose</u> considerations, and earthquakes. You start to pour wet sand out of a bucket, it doesn't pour but when you shake the bucket a bit, the sand slides out. An earthquake can cause a bad landslide if the soil is saturated (suddenly creating a low angle of repose). Big infrastructure projects take these concerns into consideration. Engineers early on developed processes for transporting corn, grains, gravels, sand, etc. down chutes from big hoppers. and into lower containers: they just attach little vibrating devices to reduce the angle of repose. It works.

Deceleration

Come to a quick stop: High speed in a kid's wagon... it hits something and suddenly stops but the kid keeps going forward ... one of Newton's Laws. Objects in motion remain in motion. Head-on car crash: people crunched into or through the windshield. Reason for seat belts, and air bags and safety sensors and braking systems in cars. You understand all that because you've been there. Just a reminder. Velocity is nice when you need it, but often it must be reduced, sometimes very quickly. Brakes (your bicycle, or heel-scrapes on your skateboard, or just skidding when running to a stop). That's <u>deceleration</u>. Trains have brakes and can reverse drive-wheel directions, big ships can reverse thrust. You know all about braking. Sometimes it's enough.

Targeting The Intersection Point

You might have noticed, or heard the stories, that some dogs, chasing a bicycle or car traveling across his field of view, will run straight toward their moving target, therefore tracing a curved path. Other dogs will run towards where they calculate their target will be when they meet: this results in a straight, shorter line. True. Some dog breeds can do that. This principle is applied in aerial combat, and in white-wing dove hunting, and in celestial mechanics. You are supposed to lead the target!

Flammability, and Thermal Mass

You build a bonfire and notice quickly that green branches refuse to burn, but dry wood catches fire readily. Certainly a crumpled-up newspaper catches fire first; that's why you use it as tinder. The explanation is that cellulose materials (trees, newspapers, etc) will burn when the temperature reached ~650 F. You understood that early on: the green branches had to dry out before the wood could get up to that temperature. Also, a big log, even though it is dry, refuses to start burning right away. It must get up to its 650F temperature, by being heated by the burning paper and tinder (small slivers of wood) around it. Applications and examples? Plenty! Look at wild-fires in times of drought. Dry brown grasses burn and spread quickly!

Fire-Redardant Materials

Houses are built with gypsum/mineral-based dry-wall siding, to keep the wood framing from burning ....to keep a little fire from spreading. Whole industries are based on application of non-combustible insulation materials, which you do understand. Asbestos makes great insulation, but it has other properties that are bad for people. Further, the available surface area is a big factor. A crumpled up newspaper weighing the same as a block of wood bursts into flame lots sooner. And a parched dry grass fire becomes a big wild fire, because the flames find a tasty easy source of fuel, compared to the fire's spread if it had only chunks of lumber (with the same dry cellulose content) to burn. Easy to understand. Insulation works.

Organoleptics

You tasted things: some bitter, some sweet, some spicy, some really nasty. You were doing organoleptics. The food and drug industries are very concerned with odor and flavor perception. The author was once on an organoleptic panel back when plastic bottles for milk and water were being developed, and when plastics additives (anti-oxidants, anti-stats, rheological modifications, etc.) were being developed. Additives leak from the plastic onto the inner surface of the bottle and therefore into the liquid. We had to describe any residual taste after the liquid had been stored in the experimental bottle, per a standard industry protocol, Great experience ... it triggered selecting additives that did not transfer flavors or odors into the stored consumer liquids (milk, water, soda-pops, etc.). One fascinating fact came out of that Monsanto Research effort. The cigar-smoking martini-quaffing executives who were in charge of the whole development effort were found to be unable to distinguish subtle flavor cues (tart, sweet, spicy, etc) and so were thrown off the test panels, leaving the million-dollar judgements (development of plastic bottle resin formulations) in the hands and taste-buds of we lowly but sensitive panelists.

Balance – Moment Arm

You have played on teeter-totters. The big kid elevates the little kid .... stranded up in the sky! But if you talk the big kid into inching forward towards the fulcrum (pivot point) or the little kid to ooch out farther toward his end, it may balance out level. That's the "moment arm" concept. <u>40 lb</u> kid at <u>3 feet</u> = (**120 foot-lbs**) balances out a <u>30 lb</u> kid out at <u>4 feet</u> = (**120 foot -lbs**). You've seen these big construction cranes suspending a heavy weight (maybe 500 lbs) in the air 40 ft out from the crane. Why doesn't it topple over? You notice that at the back of the crane structure on top there is a big black chunk of something. That's their adjustable counterbalance. It's 3000 lbs. They slide it out just enough to counterbalance that 500 lbs at 40 ft. To equalize the moment arm, they slide it out to 6.67 ft. So ..... 6.7' X 3000# = 20000 ft-lbs. 40' X 500# = 20000 ft-lbs. That's the only way to do that job. The process is typically automatic: there are strain gages and sensors that detect out-of-balance situations, and automatically adjust the position of the counterbalance weight.

Triangular Brace

You're building a "fort" out of sheets of cardboard or plywood. To keep them from falling over you brace them with sticks or poles propped up against them, forming a sort of triangle. That triangle forms a light rigid structure. Now you know why the designers of bridges, race-car roll cages, and the Notre Dame Cathedral in Paris, and your own house, etc. did what they did. It's tough to crush a triangle.

Rocket Thrust, psi

You blow up a balloon... hold it until your little sister reaches for it, then release it... the balloon takes off at high speed making a fierce blubbering noise, till it quickly deflates and plops to the floor. Why did it take off like a rocket? Exactly! There was pressure (pounds per square inch, psi) inside and a nozzle opening of a certain size (square inch). Pounds per square inch times inches equals pounds thrust. That thrust propelled the balloon until its internal pressure became zero. A rocket chamber containing the propellant doing its thing will be at 10,000 psi pressure. If the nozzle opening is 4 sq inches, the thrust will be 2500 pounds. You now know the principle behind rockets, missiles, fireworks, spinning water-sprinklers .... hundreds of applications. Plop a 50# bag of cattle feed on your foot. It doesn't hurt, but put it in a wheelbarrow and drive the wheel onto your foot: that hurts.... the wheel has few square inches, the bag has plenty; your foot feels the psi effect. You also get to know pounds per square inch when you poke a stick into mud. It goes in pretty easy. But try to shove a bucket down into the same mud with the same force. No go. Lots of square inches against the same force. Skinny stick goes in readily; big bucket goes nowhere. Grown-ups also understand the efficacy of the road builder's "sheeps-foot roller". Look it up.

Spectrum

You've seen rainbows; and have seen the rainbow spectrum thru lawn sprinkler water on a sunny day; or if you were lucky in the spray of big waves in the ocean surf; or in the mist at the base of a big waterfall. The reason is complex, having to do with water droplets' reflection, refraction, and spectral wave-lengths of light. You now know as much about it as 99% of people, including the author.

Mixed Colors

You have mixed and over-lain Crayola crayon colors ... red and blue made a sort of purple or brown, for example. You may have mixed paints or water-colors. Manufacturers of paints mix pigments of all types to create the colors that they produce and sell. That's a big industry, the pigments are carefully characterized ... particle type, size and shape, quantities, etc. to obtain the desired color.... Note that a lot of cars these days are white .... that's because TiO_2 (titanium dioxide) is a cheap white pigment. You understand their reasonings.

Stiction

You try and try, harder and harder, to pull or push or slide something free ...then suddenly the two items some apart. Stiction is a big consideration in mechanical operations. Over time, especially if there is some corrosion or chemical interaction involved, the two mating surfaces tend to grow together in a micro/nano scale. Then, under extreme stress, they break free. Workers and engineers often use shock or vibration to jostle free the stuck-together surfaces. Stiction is a big factor in major and minor dis-assembly tasks. Stiction explains earthquakes: tectonic plates are always on the move; shoving against or over or under each other; stress builds up over decades or centuries, then bam! The tectonic plates readjust their positions, and earthquakes and tsunamis cause death and destruction, and you understand what just happened!

"Kinesthetic Feedback"

You have tried to pick up something. It starts to slip. You automatically tighten your grip a bit. You carry an unbalanced tray and it tips but your hands feel it, before you do, and they make adjustments. Your hands are giving you "kinesthetic feedback". Kids' hands develop that capability early on. Robots need that capability. In the very early days of industrial robots, the "end-effectors" or grippers, were capable of great force. An early application was the handling of radioactive containers inside a containment chamber. A flask could be crushed; if the operator (safe outside the radioactive containment cell) saw it slipping, he would increase the gripping force .... the flask might break, and the cell and perhaps the whole facility would be bathed in radioactive badness! The engineers learned and developed and applied sensors and controls, so the robots could "feel" and modulate their grip, to hold and not crush the object they were holding. These "feel" sensors involved nano-scale flex/strain-gage devices. These systems are very common now in assembly and packaging operations. And your fingertips have learned how (millions of years ago) to instantly signal you that an object is sticky, or slippery, or rough, or sharp. and to modulate your grip. Tarsiers and chimps and Peregrine falcons do that easily. Takes a while for us to teach our robots.

Astronomy

The night-time view of the sky triggers most kids' interest in astronomy and celestial movement. Kids puzzle over the different times and locations of the moon. They see that the moon sometimes looks "full" and sometimes it's a "fingernail". They hear that the sun is sometimes shining right on, or that sometimes it's off to one side. They begin to understand this discussions of eclipses later on. They marvel at the Milky Way, if they're lucky to live where the night sky is clear and dark. If they don't, or have experienced a murky washed out, but cloudless sky, they begin to understand discussions of <u>light pollution, and industrial haze</u> .... and they might take steps to mitigate it, as they grow up.

Pendulums

(swing sets) are, or were, a feature of every kids' playground. Kids might not have noticed or concluded that the <u>period</u> (the total time to go back and forth), is the same whether the kid is swinging high or low. The pendulum of a big grandfather clock gets a little mechanical kick when it slows down due to friction or air resistance, but even when the swing decays or is tweaked, the period is the same, and that helps it keep time properly. Kids' experiences often help grown-up's understandings.

Thrust Vector

Controls the orientation and path of interplanetary rockets and orbiting satellites. Imagine a rocket cylinder with an engine in back pushing it, or a little red wagon with some kid pushing it. I'll go straight. But if it get pushed sideways from the back, it'll turn and go sideways in the opposite direction. Some rockets have thrust nozzles that swivel ... some have little side-pointing rockets that apply the same sideways thrust .... some ships have swiveled propellers or rudders the divert the thrust sideways... the thrust vector. When you see or hear about steering of rockets or satellites, or how to make a big ship turn; you can say "been there, done that". Note another concept you learned: if you push at the center of something, it'll just slide sideways, not spin or swivel. Some spacecraft have little rocket motors whose thrust is aligned thru the craft's CG (center-of-gravity).... which just shoves the craft sideways as intended but does not cause it to spin or tumble. Also when you see tug-boats trying to get a huge ocean liner positioned into dock-side, some of the tugs are shoving at the liner's midsection, rather than at the bow or stern, to move the liner sideways, instead of pointing it in a different direction.

Heat vs Temperature

Light a match: you get a high <u>temperature.</u> Try to heat up a big bucket of water... that requires lots of <u>heat</u>. Try to heat up that bucket with that burning match. No go. Heat is the amount, temperature is the level. The bucket of warm water has lots of heat; the match is at a much higher temperature level but very little heat. Three ice cubes can chill a glass of water nicely, but would do nothing for a big bucket of warm water. The <u>thermal mass</u> is a useful concept in these considerations. Management of heat and temperature is central to countless applications: the chef in the kitchen; petrochemical refinery processes, an aircraft carrier in combat operation, etc. Kids pick up the understanding pretty quickly if they are lucky enough to experience life.

Turbines

Your hand, out the window of a moving car..... tilt your hand and you feel the wind pushing your hand up or down. The pressure of the wind, on an angled surface, can be significant; it can generate power or motion. This effect is the basis of the function of wind turbines, your toy pin-wheel, and the control of airplanes: ailerons, rudders flaps, etc.) and much more. Hydro-electric dam power generation comes from the turbines down below being spun by the high-velocity water being released. Reversing the process ... if you provide the power (same principle) you can shove air or water to suit your purpose: airplane propellers, ships' propellers, household fans, industrial blowers, water pumps, turbo-prop aircraft, and more. They all use power to spin a propeller or fan, which provides thrust or fluid motion. Kids understand ... brushing fries away from their burgers. The air, pushed by the kid's hand, does it.

High Velocity Water

You've played with a garden hose deckled down to give a high-velocity spray. You can wash away debris or can flatten a pile of sand. Makes sense. You now understand how early gold miners did sluice mining, by directing big streams of high velocity water to bring down gold-containing dirt from hill-sides to be separated out. You also now understand a relatively new hi-tech machining process (Hydro-jet milling) using very, very high velocity jets of water to precisely cut thru thin metal parts Really .... it works!

River Meanders

You've played on little rivers or on the beach where little rivers empty into the ocean: the rivers are not straight-line: they are curvy. Why? They start out a little curvy, possibly due to stream-bed contour. At the far side of the curve the velocity is a bit higher so the river eats into the far bank; at the inner point of the curve, the velocity is lower, so sediment is deposited. Over many years, that process continues, making the curves more and more pronounced: the meanders become more extreme. On a macro scale, such as the in the Lower Rio Grande Valley in Texas, the wide meander can get cut off in a big flood, the river straightens itself out, leaving a dry or marshy crescent shaped curved feature called a resaca (former channel of a river). Check out maps of rivers, especially on the flats: they are curvy; and that's why.

Music. Resonance

You blow across the top of a bottle ....the pitch you heard is related to the volume of air being vibrated (resonated). The pitch reveals the resonance of the airflow. The slide of a trombone changes the volume of the vibrating air and therefore the pitch. Same with the valving of a cornet. Vibrations is the key. When you gently rub the wet rim of a wine glass, you hear that pitch ... when you add water to the glass, you have changed the vibrating frequency of the glass, and therefore the pitch. Most relevantly, tuning a violin ... tensioning the string ... changes its resonant frequency and therefore the pitch. Each string is, by design, a different tension and material, therefore a different pitch. Peoples' voices yield different pitches, because their vocal cords are different, but any person can change the pitch of his voice somewhat by changing the tension of his vocal cords, but of course he doesn't know he's doing that, all he knows is how to lower or increase the pitch of his voice. But now you know.

Balance. Sensors

You've done this: stand on one foot; look down at your foot and ankle. You see subtle shifts in muscles and tendons which you did not initiate, as your body maintains its balance. Your body is a complex system of automatic sensors and actuators, developed since birth. Another case: you are standing there, and decide to turn and start walking to the left .... scores of system elements automatically come into play (pressures, angles, motion, momentum, forces, etc.) as you take that first step. All you did was decide to start to walk, maybe thinking about grabbing a snack from the fridge.... your body did all the rest. When the shortstop fields the grounder and pegs the ball to second, he's thinking "do it fast and maybe we'll get a double play". His body is doing hundreds of things coordinated in nanoseconds, developed over years of practice. Engineers use strain gages, optics, pressure sensors, tilt meters, force transducers, etc. all connected with the appropriate math. Applications include systems to prevent cranes from toppling over, to optimize suspensions in race cars during the race, to guide and control missiles, etc. Nothing they create is as effective as a gymnast or ballet dancers' body systems! But you understand their efforts.

Surfaces

You learn early on that surfaces can be damaged or at least impacted. Glass can scratch plastics, and diamond-tipped tools can scratch glass. Hard metals (steel) can scratch soft metals (aluminum). You can wipe scum off some surfaces but you need Brillo or something abrasive to remove scum off other surfaces. A "non-stick" frying pan (Teflon-coated) sheds grease and carbonized debris much better than does a cast-iron pan. Some surfaces are sticky (adhesive residue, cooked egg) or slimy (oil, soap). You can abrade or even cut open human skin if you are really clumsy or clueless! The management of the chemical and physical nature of surfaces has thousands of applications across industry and folks' lives ... and you got the gist of all of that in your youth.

Behavior, "pecking order"

Kids see lots of evidence of this if they get outdoors a bit, or grow up on ranches. Hens will peck at other hens, dominant birds will swoop down and chase other birds of the same species away from the bird feeder, a horse will shove another horse aside, a big litter of puppies will quickly develop a dominance pattern, even to the point that a lower pup becomes the runt after being repeatedly shoved away from mom's food dispensing system. Chimpanzees and other primates (and elephants, Cape Buffalo, and probably even mice) constantly battle for dominance. In the wild, dominance determines breeding rights and is a Darwinian element in evolution. Dominance and pecking order in people is common, obvious, understandable and often sad.

Weather Effects

Kids experience conditions directly and personally. Cold windy sleet, hot and muggy afternoons. The seasons change. Kids understand the first two of the three concepts at play here.

1) "Conditions" describe what's happening <u>now</u> (rain, sweltering hot)
2) "Weather" describes patterns (winter is colder, spring is warmer, Florida is generally hotter, North Dakota is generally colder.
3) "Climate" is long range. The climate on Venus is hot with lots of methane.

The climate on Earth is slowly warming up because we keep burning oil and coal.

Kids (lucky enough to be surprised by a thunder/lightning storm and have to high-tail it back to the house soaking wet) understand these learned discussions because they have first-hand knowledge. They can, as grown-ups, be useful contributors to decisions on local actions (flood control levees, for instance), or global issues (legislation on fossil fuels or solar power or wind turbines, etc)

Phase Changes

You've melted an ice cube ... and have frozen water into ice cubes in the ice tray in the freezer. That's a phase change. And it happens a lot. Water vapor becomes rain (liquid) which can become hail (solid), under the right conditions. These are three maybe four phases. Liquid oxygen (a gas, typically) fuels rockets. Ammonia (NH3) is a gas, but It can be liquefied by cooling it down below -27°F. The interior of the planet Jupiter is said to include liquid hydrogen and helium, familiar gases at our typical temperature and pressures. In certain smelting processes, solid iron is heated up to become liquid, and is then cooled back to solid. Same with gold, aluminum and other metals. Phases are solid (ice, steel), liquid (water and molten steel), vapor (humidity and the oxygen and nitrogen you breathe) and plasma (a condition of ions and electrons). Plenty other examples, easily understandable. High temperature and low ambient pressure tends to drive solids into liquid and then into vapor, and beyond.

Wettability

Some surfaces don't seem to "wet". You have seen that glass doesn't wet, but wood does. Liquids roll off glass, but seem to stick to wood. True, and that's a property of the surface ... very technical, but an important property. Trying to uniformly paint a Teflon surface would be difficult, because Teflon is notorious for being non-wettable and therefore being non-paintable. Wood, or sheet-rock, or most mineral and cellulosic surfaces are wettable, therefore paintable. Wettability also correlates with ability to adhere something. Try to adhere a Teflon surface to something. Nope, you know that won't work.

Optics, Clarity, and Magnification

You see clearly <u>thru</u> some objects (transparent); but see only light/ dark thru others. That's only "translucent". Some windows are intended to provide light (translucent) but to offer a degree of privacy. Some other widows offer a clear view of the world outside. They are transparent. You've played with optical <u>lenses</u> that can make objects appear larger (convex. thicker in the center) or smaller (concave ... thinner in the center). You have probably discovered that a concave lens in bright sunlight can generate a very hot focus spot. You might have noticed that the magnified image from a big convex lens is a bit blurry around the edge ... that's spherical aberration; corrected by use of an aspheric lens. Lenses have countless applications: telescopes, microscopes, cameras. binoculars, etc. Kids learn early that that subject is complex but understandable.

Predator, Prey

Some predators lie in wait and pounce... Moray eels, Cheetahs, small-mouth bass. Others actively hunt.: lions after Thompson gazelles, dragon flies after mosquitoes, lizards after spiders, water-beetles after minnows, people after people who profess a different religion. Some other predators use camouflage to help them to try to look like the visual background, so a distracted prey morsel will wander within gobbling range. Moray eels try to look like reef rocks. Most predators use claws and/or teeth to capture their prey. Another very sharp predator (the doodle-bug or ant-lion) has evolved a cool trap: they create a sloping conical pit that ants stumble into and slip down into the ant-lion's jaws below. Some creatures (little frogs for instance) have evolved a very nasty taste and a very dramatic coloration; so that their predators learn over time and many generations that that little guy is NOT good eating. Isn't nature interesting? Kids have seen some of this and understand it.

Failure Analysis. Chain of Causes

We all learn to think critically, soon or later. If something bad happens we should learn to examine the chain of causes.

1) The tire went flat, the bike was out of the race; the favorite racer lost.
2) The wagon wheel seized up and fell off, and the wagon crashed, and the folks were hurt bad, and they didn't make it to Oregon,
3) The O-ring was selected/installed correctly but the temperature was too cold and the Challenger blew apart and crashed.

The flat tire was the result of a bad flat fix, by a worker who did not receive proper training, from a distracted manager or parent. The wagon wheel failed because of a lube job error which resulted from the lack of the proper lubricant, supplied by an incompetent supplier. The Seven astronauts died because the bosses did not believe that the engineers' concerns warranted a decision to delay the mission. Most serious failure events result from a chain of conditions and decisions. Growing up in a complex world of failures encourages us to think of root causes, not just the immediate situation.

Suspensions, Solutions

You've thrown dirt into a bucket of water. The little rocks settle to the bottom, the leaves and bits of wood float. .... but the water remains murky even after it sits around for a while. You have created a <u>suspension</u>, a mix of water and typically tiny particles with the same density and favorable surface chemistry as water. You see murky water everywhere you look. Engineers have ways to "break" a suspension, to get those things to settle out, using agglomeration and specific solvents or something (The author is a retired chemical engineer, but I must have snoozed thru that semester). A <u>solution,</u> on the other hand results when a material like salt or sugar dissolves in water, resulting in optically clear water. Of course too much of that material will exceed the water's capacity to dissolve it, leaving a goo on the bottom or your container. Note that warm water will accept and completely dissolve more solute (sugar, salt, etc.) than will cold water ... but you already knew that ...... and you understand similar situations now.

Expertise, Critical Thinking

You knew Mom was the expert, she said the broom belongs in that closet. Big brother said there might be spiders or scorpions under those slabs of lumber/plywood out back. All true. You learned slowly later that some things that people said were not true, or not true enough to be useful. You developed the capability and the necessity to apply critical thinking. Very important! If you were lucky enough to have observed and understood events, and conditions, and people, you developed the rudiments of critical thinking and reasoning. New situations and propositions require careful thought, and often more data, from appropriate sources, before conclusions and actions. (This is a vastly more important capability than any study of suspensions and solutions!)

Filtration

You've seen water spilling through sticks in a stream; or you have scooped tad-poles up out of a water-filled ditch thru your cupped fingers; or have scooped brine shrimp from a pond using your Mom's kitchen strainer. All that is crude <u>filtration</u>; removing solid material from a moving liquid. The smaller the mesh opening, the finer the solid particles that are being removed. We (the author and his little brother) partially clarified dark murky water on the Alaska North Slope tundra by pouring the water thru t-shirts. Filtration is a very big factor in a wide range of chemical and food processing operations: selection of filter screen size and media, chemical compatibility, analyses of the liquid post-filtration, replacement when filters get plugged up, power to shove the water or liquid through. All these processes and more; from petro-chem refineries, to food-prep industries, to your kitchen coffee maker; depend on appropriate filtration.

Nesting

Kids see and experience nesting in their outdoor activities ..... birds' nests are all around them, they encounter comfy little field-mice nests under old sheets of plywood. Nests are typically defended with all available resources. They see how fiercely mom blackbirds will attack people who wander too close, and how these same moms will attack and chase off that much bigger egg-eating crow. Kids learn to avoid bee-hives and wasp nests for the same reason. Wolves and Cheetahs and other beasties make dens to raise their kids, which they fiercely defend if threatened. Villages and countries are also "nests"; and those moms will send their best and bravest young men out to fight to the death to keep the neighboring tribe or country to protect their "nest". Sad but historically very common.

Materials' Elasticity

You notice that golf balls bounce and a glob of Playdoh doesn't. Golf balls are elastic and Playdoh not so much. Stretch a rubber band and it will snap back, stretch a length of yarn and it just sits there. Kids bounce rubber balls, or bounce themselves on bed-springs or trampolines. They feel and understand the concept of elasticity, before they've heard the term. They know degrees: basketball off a backboard (high).... rock off a brick wall (low), and every situation from then on. A steel ball-bearing will bounce off of the top of a rail-road track rail .... both surfaces are a bit elastic, but the ball-bearing doesn't bounce off a carpet at home: the carpet is not elastic. The property of elasticity is easily understood and is very important in life, in games, and in industry.

Heating. Radiant, Convection, Conduction

Heat is transferred by three methods:

1) <u>radiant</u> the glowing bright red elements in your toaster; or the red-hot elements in your oven on "broil" send infrared energy into your food. The food doesn't touch the red-hot elements, but receives the heat energy. Similarly, you feel the heat from a blazing fire in your fireplace, or from your campfire.

2) <u>convection</u> hot air circulates around and warms up things: the furnace in your home for instance. The warmth of the air touches the cold object. Step out into a North Dakota blizzard: cold air conducts your heat away, pretty quick.

3) <u>conduction</u> This is similar to convection, but involves contact. The warm object touches the cold object; or vice versa. You want to defrost something: you swirl it around in a big bowl of warm water. Heat is conveyed by touching. If you are cold in the Alaskan outback, you can snuggle up to a couple big sled dogs, if you're lucky. Conversely: you are tempted to pick up a slab of dry ice (frozen $CO2$ at minus 109 F): don't do it: the dry ice will conduct away your warmth leaving you with painful frostbite.

Hydroplaning

Skipping flat stones on water teaches how a speeding flat surface, angled just right creates enough lift so that the object doesn't sink. A related dangerous event happens when a speeding car's tire encounters a rain-slick road surface. It hydroplanes, losing grip and it spins out of control. When the Space Shuttle (a flat stone) re-enters the atmosphere (pool of water), it skips on the top of the atmosphere for a while until it slows down and gains aerodynamic control. Racing speedboats have specially shaped hulls and angled props so they spend much of the race just barely touching the drag-inducing water surface. Pontoon-equipped float planes speed up until the shaped pontoons lift the plane up mostly out of the drag-inducing water, so they can attain take-off speed. It's all understandable, based on your rock-skipping experience

Materials Strength. Strain Rate

Silly putty taught us a lot about strain rate. Strain rate is the speed that you strain (deform) a material. If you slowly stretch Silly Putty it will elongate and continue to elongate, but if you instead pull it apart quickly, it will snap apart cleanly. That material's tensile strength is strain rate sensitive. Other cases are familiar you have tried hard to pull something apart, but in frustration you gave it a hearty jerk and it broke! In ASTM and other engineering specification, for tensile strength, the strain rate is specified, to make possible rational comparisons of materials properties.

Materials. Thermal Expansion

Most materials (metals, glasses) grow a bit when warmed, and shrink a bit when cooled. Kids typically have little first-hand experience with this material property. One observation: washing dishes, putting one warm glass inside another, then finding later that the glasses are stuck tight together. What gives? Put an ice cube in the inner glass and warm water on the outer glass: they come apart easily. You have used the coefficient of thermal expansion, CTE. Rail-road travel: the click-click-click of the rails is much more noticeable in the winter, compared to in the summer, because the cold shrinks the rails, widening the gap between them. The gap is a necessary feature, because if there were no designed-in gap, in summer the rails would lengthen and buckle and trains would de-rail and that would be a bad thing. This property (the coefficient of thermal expansion) is very important in industry. The Alaska oil pipeline and process piping in refineries have big loops built in, so pipes can gently lengthen and shrink with the weather. Otherwise pipes would pull apart in the winter and that would be bad. All circuit boards are created with CTE compatibility so that they don't fail when they must go from hot (on) to cold (off) thousands of times. Many other examples arise in industry and construction.

Use of Tools

Kids learn about tools very early on. The tools are simple, like grabbing something and beating it against something else to make noise; or using a stick to poke your little sister. Fundamentally tools simply extend the reach or power/function of your hand (or foot in the case of swim-fins). Chimpanzees ease a smooth stick deep into a termite mound to bring out and slurp the little tasties. Some birds use little twigs to pry bugs out of cracks in tree bark. Hominids' tool use goes back millions of years. An individual's tool use (saws, lathes, drill press, hammer, etc.) and awareness of others' tool use expands with experience and situation, but the concept is well understood.

Soap Bubbles

Kids create them and love' em! What are they; how why are they formed in the shape of spheres; and why those pretty iridescent colors? The science is subtle: The soap film has slight surface tension, which keeps it from breaking into little drops. It creates a flat plane when you lift the bubble wand out and start to blow a bubble. When you start blowing air into that soapy film, the liquid soapy skin starts to stretch. This creates a surface tension or tightness in the bubble skin, and it tries to shrink the bubble into a shape with the smallest surface area to contain the volume of air in it. A sphere has that property. So that's why we have we have round bubbles. The colors are caused by the way a thin film of soapy water breaks up incident white light into a rainbow spectrum .... the colors depend on the thickness of the soap film and the light hitting a bubble at different angles causing many different inner and outer reflections. That phenomena does require some study, to really get it!

Balancing

Kids learn early on how to balance, but the decisions and actions are automatically made long before they understood what's happening. If they "feel" they are tilting or losing it, their ankles and feet and body posture automatically change to regain balance. This is an instinctive capability of primates, developed over millions of years. You are not aware of all these interactions. The tiny sensors and strain gages in our body, with their connections to all our muscles and tendons, do all the work. With practice, some of us (gymnasts and high-wire-walkers, for example, and most wild beasts) get very good at it. You can appreciate their skills.

Density

Kids understand the concepts of "density", and "heavy", and "light" before those words are learned. A lead fishing weight is "heavy". A big empty cardboard box (weighing five times as much) is "light". The math in back of the word "density" (density is weight divided by volume) comes later; and persists for their lifetime. Ocean currents are driven by the density differences in sea water that varies in salinity or temperature. Ice floats because of the expanded crystal structure of ice: it is less dense than water. Blimps and hot-air balloons rise because of the lower density of helium or hot air compared to the density of ambient air. Your toy balloon rises because it helium inside is less dense than air Easy concepts, learned surprisingly early on.

Exercise Activity

Kids very quickly saw differences in various kids' and even grown-ups' capabilities, and understood the correlations with body type and activity levels. They did not understand words like metabolism and life style and proper diet, but easily grasped and integrated those concepts and discussions in later life.

Planning

Probably one of the first attributes of an infant is planning in its simplest form. The kid wants the comfort of his personal snuggle bunny toy, so he squirms or crawls to it and settles down happily. He has imagined himself with it, and makes plans to get it, and takes action. Growing up: his bike has a flat that he must fix. He imagines the task, his plans include grabbing the wrench to remove the lug bolt, the tire-patch kit, and the tire pump. He does those things, fixes the flat (how many of use remember how to do that?), re-assembles to wheel. Pumps up the tire, and is done. Extending that planning, he further plans to procure another tire-patch kit to replace his depleted one.... his planning includes the next flat tire fix. Later he imagines and describes, as an engineering project manager on a big complex task, the totality and sequence. He specifies the materials, the procurement processes, the staffing and their duties, the facilities and equipment, the budgeting, and timing: slapping that all together on PERT and Gannt charts, and begins doing what he has imagined must be done. No different.

Materials "going bad"

Sealing things up Kids learn that you can keep a left-over half-sandwich (what kid eats only half a sandwich?) if you wrap it tight in some sort of plastic or in aluminum foil. But if you wrap it in a paper towel It dries out and goes bad. The concept of WVTR (Water Vapor Transmission Rate) comes later. Air and moisture and flavors easily race thru paper towel material (high WVTR), whereas plastic packaging films and Al-foil have very low WVTRs. That keeps out the baddies and keeps in the goodies. Kids also learn that making a good seal is critical .... a sloppy wrap job with Al-foil leaving open-air gaps defeats the purpose. A larger set of examples arise later. Sheets of well-constrained plastic keep bales of hay (legumes, alfalfa, clover) from deteriorating thru the rainy season. The food/grocery industry makes a big and necessary effort to protect food products from going bad over time. Some products are deliquescent, meaning they suck up water rapidly and must be packaged in very low WVTR packages. Also, some beverages are packaged in dark glass bottles to keep out UV light (a real baddie). These concepts serve folks in home and industry well in later discussions and decisions.

Camouflage

The outside world is a great training lab, and lucky are the kids who have access to it. Camouflage evolved to protect prey from predators. A Western Fence lizard is invisible sitting on a fence or tree or the ground, unless he wiggles. He doesn't typically wiggle but starts out a full speed when he perceives a threat. Other animals, birds beasts and snakes, have evolved camouflage coloration to be difficult to spot in the wild. Fish have evolved light color on the bottom and dark color on their top, to minimize detection from above or below. Kids, lucky enough to have spent days and years romping in local woods. are sure of this. Note, however, it works both ways: a Cheetah or leopard can disappear into local brush, waiting to pounce on a passing Thomson gazelle for lunch. Herds of Cape Buffalo will enlist a few zebras in their treks, because zebras can spot hiding lions and hyenas first, and will bolt, alerting the buff to do the same. In wartime ships are painted with irregular shapes to conceal their identity and capability; and battle vehicles are painted with camouflage coloration to hamper identification, same with uniforms worn by the soldiers. Sad necessity, but understood based on very early experience.

Ballistics

Kids learn basic ballistics almost as soon as they are able to throw things. Lob it slow and high and it will go over something. Throw it fast and level and it will go straight and hit something. They don't know the term parabola, but they know how to shape it. They also learn, later on, how to make a baseball curve. Spin it with the axis vertical, and it will curve in the spin-approaching direction: curve in or out pending on how you spin it. Throw it with the axis horizontal and it will curve up or down depending on how you spun it. Actually a hardball won't curve up, it will just not drop down as much at the predicted ballistic parabola. The batter will likely hit a pop-up if he actually hits it. Spinning it downwards (a "sinker") causes it to drop below its expected parabola. Kids learn that these curves are easier and more extreme when you are throwing a fuzzy tennis ball.... having to do with boundary layer differences. Back to ballistics: a sad but necessary application is in warfare: mortars, artillery batteries, ships' big guns .... all require sophisticated math, or just desperate pointing, to get the shell, way out there, to land where you want. Sad, but understandable. All this makes sense to kids who have ever lobbed a ball over their roof.

Rust

Kids see rust out in trash piles or playgrounds or even around the house. They know it's a bad thing; they learn later that it's a reddish-brown iron oxide formed in contact with moisture and air, over time. They learn, in order to prevent it, to slobber the exposed iron or steel items with oil or with protective paint, if the item will be exposed outside for a while. They also learn that some metallic objects appear not to rust, even without paint. Lots to learn.... and the kids are learning the necessary facts and decisions that will persist, and be expanded, for a lifetime.

Strength of Materials

Kids learn that things can be broken. You can break a flat rock if you hit it just right. You can scratch most surfaces with a jagged piece of glass. You can scratch glass with a really, really hard steel or a diamond-tipped tool. You can scrub away a lot of skin if you do a high-speed face-plant on the high-way. Kids learn the concepts of materials' strength and the vulnerability of surfaces early on, and they apply understandings and processes and necessary mitigations later in life.

Magnets

Kids play with magnets very early. The attraction, as well as that mysterious N–N or S–S repulsion fascinates them. They learn later about great big cranes with big magnets that lift squashed cars in junk yards, or transport heavy iron/steel objects (boilers, wheel/ axle assemblies, frames etc.) in metal smelting or processing, or in large-scale assembly operations. The applications are new but kids understand the principles. They hear about permanent vs electro-magnets, and understand the operation of electric motors (sequenced timed pulses of electricity into rotors and stators within an armature ... it's complex but understandable to kids who have played with magnets. One note of serious caution to kids and grown-ups: do NOT let infants play with and swallow little magnets. Those magnets can lodge in different parts of an infant's intestinal system, and attract each other clamping together and perforating the tissues; leading to fatal consequence.

Territoriality

Kids get a hint of territoriality and personal space, even in its simplest form, at a very early age. Before words are available: "this is MY end of the sand-box". "stop following me around". Later they learn about the pride of lions defending their turf, or about prehistoric villagers (trying to protect its desirable fishing spot or grove of local fruit trees) fighting an encroaching neighboring tribe. World wars and millions of deaths have been caused by large scale turf wars. People, sadly, know the feeling.

Experimentation

Kids are not issued instruction manuals, but they learn very early how to experiment, and they do so with almost scientific rigor. Something fails or doesn't work right. You try something (one thing) different; and see the result, (Look up 1FAT, DOE). Later they learn that sometimes things work and sometimes thing don't work; and that there are often other factors at play. They either explore what else might be operative, or they learn the gist of the likely-hood of something happening. Welcome to the world of probabilities and sophisticated statistical math; Kids learn early on that things are not always exactly as they seem, or do not always work out right, and that careful experimentation is necessary, or that they must accept some uncertainty. Later they can understand and can contribute to efforts by specialists, exploring important issues to apply proper experimental protocols to arrive at desired results.

Gag Reflex

Kids, from day one, spit out nasty tastes. This is a Darwinian survival reflex (most organisms, in our case vertebrates and hominids) develop the instinct that bad tasting things are typically poisonous. Over time, kids graduate from breast milk to Gerber Baby Foods, then to people food; but many of these new flavors trigger the reflex to spit it out. That's natural. If the new food gets too far down, the gag reflex comes into play. But kids do get used to the flavors of people food. Grown-ups certainly develop taste preferences, but this is far from Darwinian survival: "you gotta develop a taste for it" or "I really love Mexican foods, or Thai cuisine, etc".)

Tides and Waves

Lucky kids grow up near a lake or the ocean, and experience some important phenomena. Throw a big rock in the lake and the ripples radiate out and wash up onto the shore. A tsunami is the same thing, caused by some big underwater event far away out in the deep ocean, like an earthquake or a sudden tectonic shift, causing big ripples in the ocean, which can cause havoc when they hit our coastline. Also, kids see waves being created by strong winds sweeping up against the shore. Hurricanes do the same thing and worse: water level shoved up against the shore by the wind, plus huge waves on top of that. Add in a high tide (some times and places that could be many feet). Combine all that and you understand a hurricane's "storm surge". Another factor in storm surge is the reduced barometric pressure in the center of a hurricane, raising local ocean level by a few inches, but kids and grown-ups can't feel that effect. But kids do learn enough in their early years to understand storm surge discussions and even predictions, later on.

Taking Things Apart

Is a favorite kids' activity. Putting things back together, maybe even fixed and better, is also familiar. The tasks and challenges get bigger and more complex, and they learn the basics of friction, tolerances, wear and deterioration, tools, sequences, cause and effect, diagnostics, and more, and even the decisions to fix vs. scrap/salvage. For some kids that leads to interest in and skills preparing for satisfying careers in engineering or mechanics and beyond.

Cuts and Bruises

Are every day in the life of most kids. They learn causes, effects, seriousness, probabilities, and prevention (basics in engineering). The learn about infections (deep science behind this subject!), trips to the Emergency Room, bandages, casts and splints, and are becoming prepared for a life of action, possibly a career in medicine, as well as, more importantly, for the role of good parenting.

Pets

Lucky kids grow up with pets .... puppies, kittens, gerbils, goldfish, even little turtles, lizards and more. They learn the responsibility of having living creatures depending on you, of feeding and environment; and the sensitivity to the needs of other creatures; people included. Lifetime lessons, not really engineering science, but the fundamentals overlap.

Math

Kids count things before they know they are counting. The see a pile of many things, then they know that there are fewer things after some are removed. They see kids adding more things to the pile ... more and a bigger pile. They see eight things divided equally amongst four kids, or something similar. They learn numbers, and that numbers mean something, and that numbers can be bounced against other numbers to give you answers .... to solve problems. That seems like the definition of mathematics, and most kids catch on quick.

Later

Kids who have been lucky enough to have experienced and to understand these things, and have learned to think critically and to seek explanations, will be able later to deal with new phenomena and concepts, such as: Why are natural sandstone arches in the wild west and roof supports in cathedrals shaped like catenaries (the shape of a dangling chain held at both end at the same level). What is birefringence, and Moire´ patterns, and iridescence? Why does the shape of some organic molecules cause them to react much differently than other organic molecules that are chemically identical? They can contribute to the questions (this is important!) as well as to the answers.

Final Comments

Where does technology begin and when do certain development take off? Back-street shops often create solutions and materials that later are discovered or expanded by main-stream technologists. Aerospace folks had much to learn from skate-board builders. The hula-hoop and the big-wheel craze in the 60s created million dollar HDPE polymerization units filling that demand, which enabled growth of milk and bleach and soap bottle and molded gas-tanks. Polyurethane wheels for skate-boards triggered development of urethane chemistries that are the backbone of current industry and commerce. Hallicrafters crystal sets were created to tease out a spark of radio signal, which sparked a generation of geeks, who created our 21st century technologies. Surf-board fab trial-and-error spawned a generation of engineers who translated that experience into DOE disciplines that they included in experiments at the Thiokol Propulsion facility. REI tents inspired space-craft designers to conceptualize self-unfurling solar-sail structures. Don Garlits and his dragster cohort hot-rodders squeezed mega-HP out of flat-heads and early hemis by manifolding and co-tuning exhausts and intake, finessing cams and ignition, while Detroit was building anchors. Teen-age gamers, early in the D&D years of role-playing began writing their own code and developing eye-hand and situational awareness skills; and now are creating the simulation and guidance systems for, and becoming, the war-fighters of the 21st century. NASCAR and Formula racers equipped cars with inertial and proximity and strain-gage sensors 20 years ago, for real-time performance optimization and even closed-loop servo; now being embodied in mainstream passenger cars and trucks. Science-fiction writers imagined mini/nano devices w integrated communications, suggesting feasibility and utility long before we created MEMS and 2.5-D packaging. And lest we forget the power of <u>market-pull</u>: Our grand-daughters now, with no fear of computers and keyboards, and with the expectation of instant gratification and universal communication and on-demand information and entertainment, continue to pull products and technologies into existence. And not to mention the ancient pyramid builders and Damascene-sword-makers skills; how their skills and accomplishments continue to inspire engineers and astronomers and metallurgists today. Inventors don't start when they are 40 years old... they start thinking things out when they are five years old! Big developments are often the result of something that that kid did not find satisfying years ago.

Afterword

This has been a fun exercise; trying to help others realize that they know more than they think about engineering science, because of what they experienced in all their activities when they were typical kids. I started this book some twenty years ago. However, lately, I've come to fear that it's becoming irrelevant. That's because many kids these days grow up only squinting down at some little digital device, playing artificial video games and texting irrelevancies to a kid somewhere else in the world doing the same thing. They have never chased a little garter snake, or spun themselves off into the dirt from a play-ground merry-go-round, or crashed a mountain bike, or fixed a flat, or watched the Pleiades, or waded a stream to try to catch minnows. Their new skills may equip them to contribute into the 21st century, whereas my kids' skills are equipped to understand and appreciate the last century, rather than the next. Still, there is much in this book, I'm hoping, that will connect with some folks and kids today.

Printed in the United States
by Baker & Taylor Publisher Services